A. M. Al-Mukhtar

Spot Weldabaility Principles and Considerations

A. M. Al-Mukhtar

Spot Weldabaility Principles and Considerations

Welding procedure and parameters Quality of the welds

Südwestdeutscher Verlag für Hochschulschriften

Imprint
Any brand names and product names mentioned in this book are subject to trademark, brand or patent protection and are trademarks or registered trademarks of their respective holders. The use of brand names, product names, common names, trade names, product descriptions etc. even without a particular marking in this work is in no way to be construed to mean that such names may be regarded as unrestricted in respect of trademark and brand protection legislation and could thus be used by anyone.

Cover image: www.ingimage.com

Publisher:
Südwestdeutscher Verlag für Hochschulschriften
is a trademark of
International Book Market Service Ltd., member of OmniScriptum Publishing Group
17 Meldrum Street, Beau Bassin 71504, Mauritius

Printed at: see last page
ISBN: 978-3-8381-5129-8

Copyright © A. M. Al-Mukhtar
Copyright © 2015 International Book Market Service Ltd., member of OmniScriptum Publishing Group
All rights reserved. Beau Bassin 2015

Spot Weldabaility
Principles and Considerations

Welding procedure and parameters
Qualität von Schweißverbindungen

A.M. Al-Mukhtar

To Dania and Bainat

Dr.-Ing. A. Almukhtar

Assistant Professor, Department of Mechanical Engineering
College of Engineering
Dhofar University
Salalah, Sultanate of Oman

Technische Universität Bergakademie Freiberg
Faculty of Geosciences, Geoengineering and Mining
Institute for Geology, 09599 Freiberg, Germany

Department of Automated Manufacturing Engineering
Al-Khwarizmi College of Engineering, Baghdad University,
Baghdad, Iraq

Table of Contents

SPOT WELDABILITY 5
Principles and Considerations 5
Preface 5
Chapter 1 6
Joining Technology 6
   1.1.   Introduction 6
   1.2.   Types of Welding Processes 8
       1.2.1.   Fusion Welding 8
       1.2.2.   Solid State Welding 8
   1.3.   Resistance Spot Welding 9
       1.3.1.   Spot Welding Equipments 10
       1.3.2.   Resistance Welding Electrodes 12
       1.3.3.   Lobe Curve 14
       1.3.4.   Applications of Welding 14
   1.4.   Austenitic Stainless Steel 15
Chapter Two 17
Spot Welding Process and Experiments 17
   2.1.   Introduction 17
   2.2.   Experimental Analysis 17
   2.2.1.   Specimen Preparation.. 18
   2.2.2.   Measurements 18
   2.3.   Joint Strength and Fracturing 19
   2.4.   Determination of Weldability 21
Chapter 3... 23
Strength Evaluation of Spot Weld 23
   3.1.   Introduction 23
   3.2.   Spot Welding of Carbon Steel Sheet 23
       3.2.1.   Static Test 24
       3.2.2.   Welding Equipments 25
       3.2.3.   Welding Process 25
       3.2.4.   Tensile and Shearing Testing 26
       3.2.5.   Determination of Weldability 26
   3.3.   Results and Discussion 27
       3.3.1.   Effect of Weld Current and Weld Time, Lap Joints 27
       3.3.2.   Effect of Weld Current and Weld Time, Peel Joints 29
   3.4.   Weldability Lobe Curve 30
   3.5.   Conclusions 31
Chapter 4... 33
Cracking in Spot Welded Joints 33
   4.1.   Introduction 33
   4.2.   Welding Cracks 34
   4.3.   Stainless Steel Spot Weld Structure 36
   4.3.1   High Temperature Crack Growth 36
   4.3.2   Microhardness Distributions 37
   4.3.3   High Temperature Crack Growth and Ferrite Contents 38
   4.3.   Spot Welding Defects.. 42
   4.4.   Conclusions 43
References 44
Index 49

SPOT WELDABILITY
Principles and Considerations

Preface

SPOT WELDABILITY, Principles and Considerations work introduces the basic spot welding principles. It presents the overview about the mechanism, experiments, limitation and defects of this process. In this edition, new concepts, awareness were presented.

The results from the series of scientific works and literature are discussed. The welding parameter effects and welding metallurgical aspects are presented. The text covers four Chapters. Chapter 1 presented the overview about the welding technique, with focusing on the spot welding as the aim of this text. An overview of spot welding, experiments and the properties of the produced welds were presented in Chapter 2. Moreover, the metal's spot weldability; strength and fracture are the main topics on the spot structural analysis was discussed in Chapter 3. The studying of the welding variables effect on the mechanical properties of the structures is the key analysis. Therefore, the effect of welding parameters and conditions are discussed. These considerations are initiated due to the real need in the industrial application. The applications of the normal industrial forms of spot welding require an acceptable welding that withstands high pressure and temperature that exists in some application. Several studies show that the welding variable has an effect on the hardness of spot nugget and on the cracking. Chapter 4 shows that the heat treatments (annealing and stress relieving) increase the softening due to the stress relieving from the spot nugget. A metallurgical examination is carried out made for weld area, including cracking and phases. The crack growth occurring due to the holding time in the temperature of 750 °C. The delta-ferrite amount increases with the heat input as confirmed by the magnetic scope examination. The most common defects that could experimentally appear in spot welding were presented. Therefore, the fracture ability and the microstructureal examination were presented in Chapter 4.

2015   Dr.-Ing. A. M. Al Mukhtar

Chapter 1
Joining Technology
1.1. Introduction

Welding is material joining processes in which two or more parts are coalesced by a suitable application of heat and/or pressure. The history of forced joining which called today welding; is extended into the historians for many thousands of years about 3000 B.C. At this time, a low energy was available. Hence, the joining copper, gold, silver, and lead-tin alloys were carried out only, see Fig. 1.

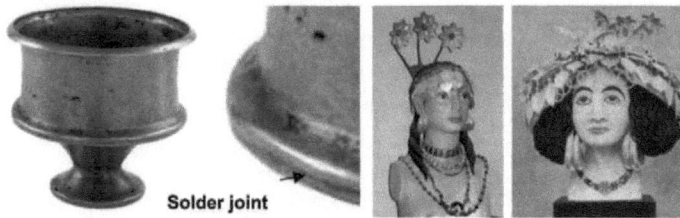

Figure 1 Silver (Ag) and Gold (Au) Electrum cup and golden headdress kings grave at Ur (Mesopotamia / today Iraq), 2600 B.C. [1].

For the low melting metals, the joining process was limited to soldering, brazing, and forging, see Fig. 2 [1].

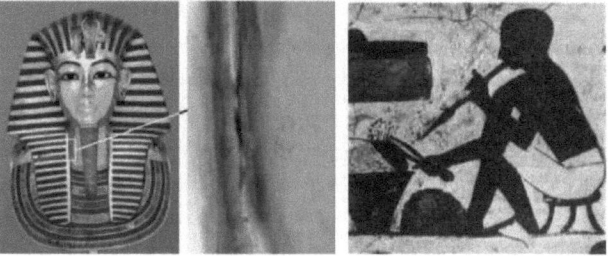

Figure 2 Soldering and forging [1]

Development of modern welding technology began in the second half of the nineteenth century between 1880 and 1900 because the electrical energy became available. Two men who developed the primary concepts of resistance welding (electrical welding), J.P. Joule and Elihu Thomson in England in 1857, and the process brought to public notice in 1886 for the first time [2]. It was Joule who discovered the relationship $H = I^2 RtK$ where H is the total generated heat in watt-seconds, I is the current in amperes, and t is the time in seconds. R is the resistance through the material in Ohms, and K is a factor representing heat losses [2].

The three welding processes, namely; arc welding, resistance welding and Oxy-fuel welding; constitute by far the majority of welding operations performed today. Today, some new techniques have developed with the same concepts that were developed before two centuries. Many welding processes are accomplished by heat alone, with no pressure applied. Other processes carried out by a combination of heat and pressure, and still others by pressure alone with no external heat supplied. In some welding processes a filler material is added to facilitate the coalescence.

Welding is most commonly associated with the homogeneous metal parts; however, the dissimilar metals are welded today. The assembly of parts that are joined by welding is called a weldment. Engineering structures and components can be constructed by compiling different methods (i.e., Riveting, bolting, and welding). These parts could be manufactured from the traditional forming processes. The welded structures have advantages over the riveted structure due to their efficiency and tightness, e.g. in containers and storages of water and air. No additional weight or materials are added, hence weight saving. The designers present a list of design relations and safety factors for riveted and bolted joints. These factors are related to the sheet thickness and dimensions. In welded joints, there are no limits on thickness. Therefore, the design is simple in welding. The required time and cost for fabrication is also reduced. Welding provides a permanent joint. The welded parts become a single entity. The welded joint can be stronger than the parent materials if filler metal is used that has strength properties superior to those of the parent. No additional materials and pieces are required in term of fabrication and costs as compared with riveting, and bolting. The latter is heavier than corresponding weldments, see Fig. 3.

Figure 3 Bridge structures with tousendas of rivets and bolts

Welding is not local restricted. It can be accomplished in the field. Welding can be mechanized and permits considerable freedom in design [2].

On the other hand, the welded joints, suffer from the effect of metallurgical, and geometrical discontinuities, and cracks. Moreover, most welding operations are performed manually and are expensive in terms of labor cost. The operations are considered skilled trades. The using of high energy is inherently dangerous, see Fig. 4. Welding produces a permanent bond between the components, hence, it does not allow for convenient disassembly, then welding should not be used as the assembly method. The welded joint can suffer from certain quality defects that are difficult to detect. The defect can reduce the strength of the joint. A welded joint, for many reasons, needs stress relief heat treatment [2].

Figure 4 Welding cracks

1.2. Types of Welding Processes

Different types of welding operations have been categorized by the American Welding Society (AWS) according to the power and using of energy, see [3]. Hence, the welding process can be divided into two major groups; fusion and solid state welding as follows [2][4].

1.2.1. Fusion Welding

These processes use the heat to melt the metals. In many fusion welding operations, a filler metal has to be added to the molten pool to produce the singularity of the welded joint. The most widely used welding processes include the following groups [2][4]:

a) Arc welding (AW).
b) Resistance welding (RW).
c) Oxyful gas welding (OFW).
d) Laser beam and electron beam welding.

1.2.2. Solid State Welding

It refers to the joining processes in which coalescence results from the application of pressure only [5]. The combination of heat and pressure also considered as a solid state. But the used temperature is below the melting point of the metals being welded under the pressure. No filler metal is utilized in solid state processes such as:

a) Diffusion welding (DFW).
b) Friction Welding (FRW).
c) Ultrasonic Welding (USW).
d) Impact welding.

1.3. Resistance Spot Welding

Resistance welding (RW) is a group of fusion welding processes. It uses a combination of heat and pressure. The heat is generated due to electrical resistance to the current flow, hence producing what's called weld nugget between the two surfaces. The principal components in resistance welding are shown in Fig. 5.

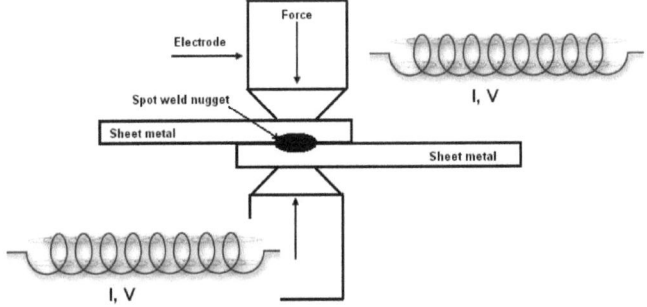

Figure 5 **The components in spot welding process**

The components include work parts to be welded (usually sheet-metal parts), two opposing electrode contact overlapped sheets, a means of applying pressure, and an AC power supply from which a controlled current can be applied. As compared with the arc welding, resistance welding uses no shielding gases, flux, or filler metal. Moreover, the electrodes that conduct electrical power to the process are non-consumable. Therefore, the initial cost of the machine is high. The lowest skill and good reliability are the main advantages as compare with arc welding. Nevertheless, tip machining is required from time to time to maintain the dimensions and maintain the current density.

The required heat is generated through the electrical resistance between the two sheets to be joined. Therefore, in the spot welding, welding current, electrode force, welding time and holding time were considered as welding effecting parameters [6].

The weld nugget area is influenced by the heat energy which required to melt a weight unit of materials. Therefore, the density and the conductivity of materials should be considered to melt the weld nugget. Therefore, five methods of resistance welding namely; spot, seam, projection, flash and upset welding. The lap joints are used in the first three processes and the butt joints in the last two in which considered as a limitation, see Fig. 6. An oxide film, altering resistance. Therefore, the surface preparation is required. The heat generated is followed by the application of pressure to ensure a good bond. The accurate control and timing of the electrical current and the application of the pressure is an essential feature in resistance welding [7].

Figure 6 **Lap joint; and butt joint**

Welding may be performed by means of single or multiple electrodes to ensure the required pressure supplied through mechanical or pneumatic means [5]. The power supply generally alternates current. The timing of the current application is typically measured in terms of number of cycles. Parts inserted between open electrodes, hence electrodes is closed and force has applied. The weld time and current are switched on. After a certain time, the current is turned off, but force is maintained or increased. Sometimes, a reduced current is applied to enhance the weld metal distribution, see Fig. 8.

Figure 7 Airplane spot welded sheet wing, Boing 737.

1.3.1. Spot Welding Equipments

The quality requirements of the joint geometry and the production time are leading to select the resistance welding equipment.

Complex resistance welding equipments may be used for some applications, hence a special type of electrodes design and construction have to be used. The standard pieces of equipment have three principal elements as follows, see Fig. 9 [5]: -

a) The electrical circuit: it consists of a transformer, a current regulator (tap charger) and a secondary circuit. All these parts conduct the welding current to the electrodes.
b) The mechanical system: this system holds the work-piece and applies the required welding force or pressure.
c) The control or timing apparatus: these controls may sometimes be complicated, and affected by other variables; in the simple machine the control of time only will be considered.

These equipment elements control the three principal parameters; the current, the force or pressure, and the time.

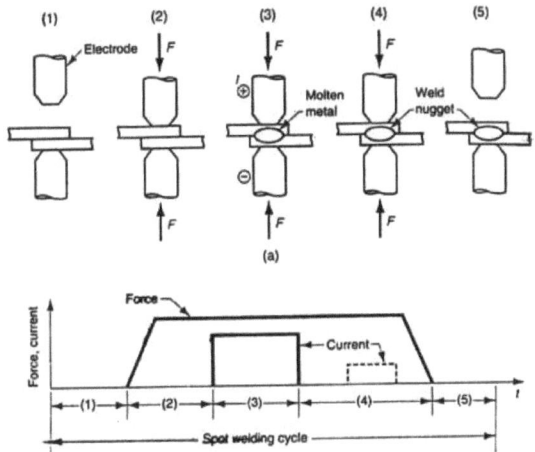

Figure 8 (a) Spot welding cycle (b) force and current during the cycle

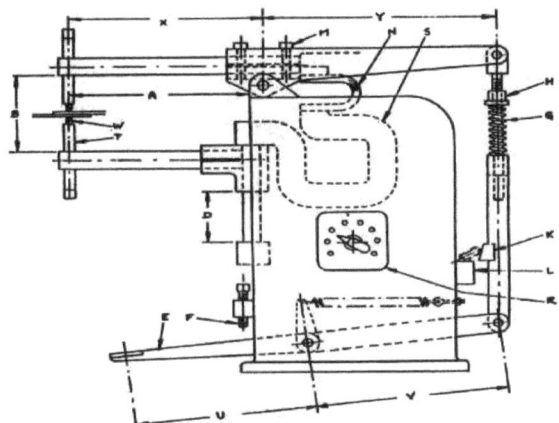

Figure 9 Machin parts: A, Throat depth. B, horn spacing. C, centerline of horn lever trunnion. D, lower arm vertical adjustment. E, foot lever adjustable which regulates the amount of electrode opening. G, welding pressure spring. H, welding pressure spring adjustment which regulates the amount of welding pressure. K, initiating switch cam or trigger. L, initiating switch. M, horn cap screws, N, secondary flexible bands. R, current regulator. S, transformer secondary. T, water-cooled electrode holders. W, welding points or electrodes [8] .

1.3.2. Resistance Welding Electrodes

The selection of the proper electrode for a given resistance welding application can be based upon metal type, surface conditions, applications and product type.

Resistance welding electrodes perform a number of electrical and mechanical functions as follows [8]:

a) They transmit the welding current within a restricted cross section. Therefore, they determine the current density per unit area of electrode tip.
b) The electrodes are also used to apply pressure over the area of the weld to forge the molten area simultaneously. The nugget area will be forged when the area has a proper heat temperature.
c) The electrodes are machined to have cooling channels inside to prevent the overheating and surface fusion of the work-piece.
d) The electrode pressure is also employed to hold the workpiece in proper alignment (act as welding fixture).

To avoid the defects in the electrodes and in the weld, the electrode material should have a compressive strength, hardness and electrical conductivity. Commercially pure copper is an excellent electrical conductor. The availability of other material which have adequate electrical conductivity together with superior mechanical properties, reducing the using of pure copper.

The compressive strength of copper is low. Therefore, it anneals at a very low temperature, causing it to wear very rapidly. Materials used for resistance welding electrodes must have the proper properties that are satisfied with a work piece. In general, relatively high-conductivity electrodes should be used to weld low-conductivity materials. In other hand, low-conductivity electrodes should be used on high-conductivity materials. To determine the electrode durability (thermal fatigue), trial and error is used. Table 1 shows the electrode materials and properties.

Table 1 Electrode materials [3]

Material	Electrical Conductivity %	Brinell Hardness	Max. Working Temperature °C
Pure copper	98	80	150
Copper, 1% cadmium (cold drawn)	85	100	200
Copper,1% silver (cold drawn)	95	90	250
Copper, 2.5% cobalt, 0.4% Beryllium (heat treated)	55	200	400
50% copper, 50% Tungsten (sintered powder)	40	110	Above 600
Pure Tungsten	25	250	Above 700

[1] ***Cu-Cr alloys** that contains 0.4-0.6% Cr have electrical conductivity 80% of pure copper and 155 Brinell hardness after heat treatment by quenching and rehearing.*

The design of resistance welding electrodes depends upon many factors such as accessibility of weld area, composition and thickness of the parts being welded, see Fig. 10. The surface conditions also play a significant role. Because the alloying with the electrode materials. Moreover, principal consideration must be given to the electrode material, electrode geometry (dimensions) and cooling facility.

To distribute the squeeze and compression load over the weld area, the ideal shape for spot welding is the straight faces electrode. It allows more fixed capture for the pieces. Therefore, the electrode tip wear is more fairly uniform and keep lower. However, flat faces electrode use when the weld area is in accessible regions.

There have been proposed other suitable shapes of electrode to allow the accessible to the weld area. Some typical spot welding electrodes are illustrated in Fig. 10 [5].

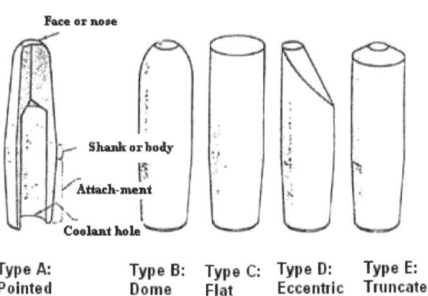

Type A: Pointed Type B: Dome Type C: Flat Type D: Eccentric Type E: Truncate

Figure 10 Typical welding electrodes [3]

1.3.3. Lobe Curve

Lobe curve is used to announce the spot weldability [7]. For a given combination of variables (material, pressure, etc.) a range of current and welding times will produce acceptable welds, [9], [10], see Fig. 11. On this matter, many articles have been published in termed of spot weldability. If the welding time is insufficient for the certain current, the nuggets will not have adequate time to grow. Therefore, a brittle weld will be result on the left of the curve. By using the proper time, nuggets of adequate strength will be produced in the middle region. The spot welds Lobe curve can be found by systematic testing by specifying a distributed range of current values and testing weld made with varying welding times. These curves are valid for specific combinations of welding conditions. The type of metal, surface preparation and roughness, workpiece thickness and electrode tip pressure are among the factors that affect the curve. Changing one or more of these factors may change the curve width or shift its position. Moreover, the acceptable regions have a variety of weld size and strength.

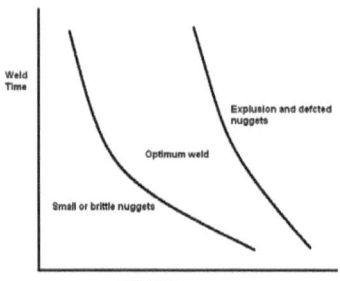

Figure 11 Lobe curve

1.3.4. Applications of Welding

In general, the welded joints have a broad applicability in the car industry, aeronautics, pipelines, and structures [11]. Resistance welding is a flexible method for joining the metals. Because the fixed and portable arms can be used. In addition to the wide range of materials, which the machine can be automated accordingly.

The most useful application of R.W are aircraft, fuselage, landing gear, exhausts rings, wing assemblies, tail assemblies and wheels. Nevertheless, R.S.W using in the jet engine manufacture such as combustion chamber and an outer shell, see Ref. [5].

Spot welds produced via resistance welding have been widely used in the joining of sheet metal for auto bodies since the 1950s. Every modern car contains over 2000 spot welds [12].

The modern automobile is essentially a resistance-welded structure made up of thousands of individual works. The R.S.W is widely used in mass production of automobile appliances, metal furniture, and products made of sheet metal. The resistance spot welds (RSW) process has attracted attention to the fabrication of aluminum structures in automobile manufacturing due to the weight reduction [13].

If one considers that a typical car body has approximately 10.000 individual spot welds and that the annual production of automobiles throughout the world is measured in tens of millions of units, the economic importance of R.S.W is used can be appreciated [5]. In addition, the air vehicle industry is using the spot weld in many parts. The building construction, communications equipment, telephone and electronic equipment.

1.4. Austenitic Stainless Steel

Most studies investigated the welding of the low carbon steel, aluminum and high strength low alloys steel due to their wide application in automotive industries. Little investigations deal with the spot weldability ofit'sainless steel because its used with the resistance spot welding [14]. The products which suffer both corrosion and working in sever conditions are made from austenitic stainless steel. This metal has excellent mechanical properties and excessive corrosion resistance. In any application, the weld region subjected to severe the conditions. The working at temperature up to 900 °C and pressure will produce cracking [14]. Especially when corrosive fluids are exist. Therefore, the determination of the optimum welding parameters that develops strong weld to prevent any failure during service time is required. Therefore, a set of variables is suggested to get the behaviour of the weld strength and weld area with the welding variable (weld time, current, and electrode pressure). Therefore, the heat treatment such as annealing and stress relieving, are preferable for the weldments properties and its microstructure.

The gas turbine engines of aircraft and rockets have stainless steel components. They are welded using the RSW. These components will undergo excessive service operations like temperature, pressure and corrosive media. Thus, stainless steel offers significant advantages over the other materials because of its high strength, toughness, corrosion resistance and ease of fabrication, see Fig. 12.

Figure 12 Corrosion resistance of Austentic stainless steel

Chapter Two
Spot Welding Process and Experiments

2.1. Introduction

Most investigations aim to assign the proper combination between welding parameters and the producing of the sound spot welding. To study the influences of welding conditions and to determine the effective welding variables, the influencing variables such as current, time, electrode pressure, and surface condition have to be assigned. Each of these variables affects the heat generation. The relationships between the affective variables on the joint strength and weld area, weld current, weld time and electrode tip diameter (electrode pressure) were determined.

2.2. Experimental Analysis

The experiments show several types of tests that determine the spot weldability of metal. However, the materials being welding play a significant role in the welding machine selection, traditional machine is used widely, see Fig. 13. Austenitic stainless steel (321) and 304L were investigated, see Refs [5] [15] [16] [14]. Nevertheless, other materials like carbon steel, aluminum alloys and titanium are welded widely using RSW [7] [17] [18][19][20] [21] [22] [23].

The effect of post weld heat treatment (PWHT) on the mechanical properties of spot welded joints was included in different studies [24]. The following procedures are usually carried out [14], [25]:

1- The sets of peel and lap specimens are made under different welding conditions (weld time, current and pressure).
2- The static tensile test is made to determine weld serviceability. A number of experimental results have been presented. The relationship of weld strength (maximum load) and weld nugget area with welding time and current.
3- The Lobe curve can be determined for those specimens.
4- The cross section is made at the center of the spot weld for the as welded specimens and for PWHT specimens. This is usually made to correlate the metallurgical factors with the mechanical properties. Moreover, the number of defects was investigated.
5- Micro-hardness tests are applied along the weld nugget.

The modern spot welding machine has developed due the industry requirements. There are many various kinds of spot-welding devices, sometimes referred to as resistance spot welding systems, which are suitable welding operations. Machines with a pneumatic cylinder attached, widely used [26].

Figure 13 Rocker–arm spot welding machine.

2.2.1. Specimen Preparation

The two geometries were used widely in mechanical testing according to AWS as follows [8][27]:

a) ***Lap specimen:*** The two flat strips joined in an overlapping section. For the 1.5 *mm* thick plates were (35x120) [14]. The geometry was approximated as closely as possible to have pure shear mode on the joint. A major concern is to maintain symmetrical loading in the middle of the overlapping distance, to prevent any moment around the weld nugget specimen.

b) ***Peel specimen:*** The peel mode specimen was subjected to a combined stress created by joining the specimen with same distance away from the loading plane. This was done by putting a bend in the plates. The stress was a tensile load combined with a moment, which was dependent on the distance of the load plane to the weld joint, as the joint got closer to the loading plane, the behavior was closely approximated pure tension. A 90° bend was then made. According to the assigned standard the overlapping area is depending on the sheet thickness.

Burrs should be removed from the overlapping edge of components being resistance welded. Unremovable burrs causing the current shunting through the burr instead of passing through the contact point (weld nugget). This will reduce the current that required to produce the proposed nugget area.

2.2.2. Measurements

The following measurements were performed as follows (American Welding Society (AWS) [3]:

1. ***Primary current***: One of the most reliable methods of measuring the primary current is by means of an indicating pointer clamp meter.
2. ***Secondary current***: The most commonly accepted method of measuring the secondary current is to multiply primary current by the turn ratio. (Line voltage divided by open circuit secondary voltage provides the turn ratio for any type). The values of primary and secondary currents are limited by the capacity of welding machine. The current increases to the point where metal expulsion occurs. Optimum

strength may thus be obtained. The open circuit secondary voltage is measured by the two methods, the element number 2 (voltmeter), and element 3 (clamp meter).
3. **Time**: compute using Electronic timer that is connected to the electrical contractors. The minimum range of timers is 15 cycles. The value of times shown in this work is in the cycle (1 cycle=1/60 second). The timer is shown as element number 4.
4. **Pressure**: is calculated using the most common method of determining pressure as follows:
Electrode force = foot force× [(u/v) × (y/x)], where, u=45 *cm*. v=59 *cm*. y=69 *cm* and x=17 *cm*.

According to the machine design, this ratio approximately equal to (3) which is confirmed by the AWS recommendations [8]. The foot force can be measured and determine the electrode pressure on the nugget.

2.3. Joint Strength and Fracturing

Static strength was measured using conventional tensile and shear test specimens. Specimen details and testing method were made according to the AWS practice [7].

The tensile shear using a conventional hydraulic tensile testing machine was equipped with a digital readout for load and extension. The peak load and break extension for each type specimen has to be recorded simultaneously [7]. Thus, the maximum load was recorded as a measure of the weld strength [25]. The chart speed set to be suitable for each type of specimen usually varies from (2 - 10 *mm/min*), and depend upon the welding condition.

The interfacial fracture of lap-shear specimens is at the faying surfaces, due to the low twisting moment, and due to the shear stress effect [14].

The conventional peel test was performed also in tension. At high welding settings, the failure of the peel test is combined between the interfacial fracture into the weld metal and pullout fracture [14], Fig. 14.

The failure modes of resistance spot welded thin plate structure, pullout and interfacial failure modes, were investigated based on thickness of joining plates and load secondary voltage to determine the critical nugget diameter [28]. It was found that the weld nugget of 2-1 mm and 3-1 mm specimens were asymmetric form. That was the expected results, since the symmetrical nugget is produced for similar materials and thickness.

At a lower welding setting, the small or brittle nugget will be produced [10]. Thus, the failure will be interfacing at the faying surface. The weld button diameter was measured using a caliper gauge, vernier caliper.

The fracture area is measured along the weld nugget interface without the expulsion of the molten weld metal. The weld metal fracture surface of Mn-Mo-Cb steel was identified by its bright appearance and surrounded by a halo having a gray appearance [29], see Fig. 14. The high ductility causing pullout of the weld metal at the center of the spot, see Figure 14 (Right).

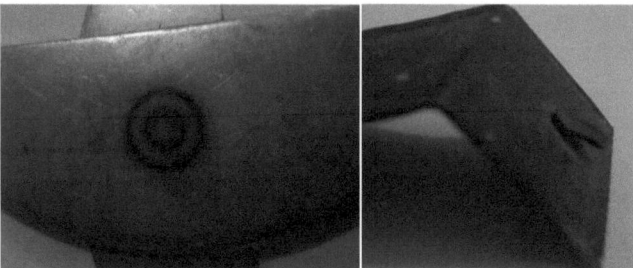

Figure 14 weld nugget appearance and plug fracture

The fracture surfaces and failure mode have investigated in different studies. The has been observed interface and plug and hole type in carbon steel and aluminum alloys [30], see Figure 14.

Figure 15 Failure mode of M190 grade steel after peel test: interfaceilure and (b) interfacial failure

It is necessary to determine the hardness over a small area of material. Therefore, micro hardness tests are employed to examine the hardness variations along the weld. Both traverse and longitudinal directions are investigated due to the directional properties [14], see Fig. 16.

Vickers hardness is employed, using a conventional microhardness tester (JTT Digital micrometer taster, type JMT7 type A, Toshi INC.) with (300 gr) load [14].

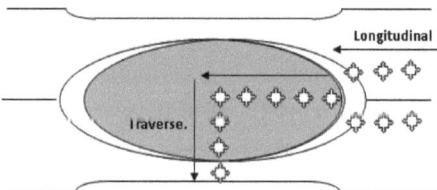

Figure 16 The sequence and direction of microhardness measurements

The same specimens of microhardness test were prepared for the metallurgical tests to reveal the effects under welding condition and after a heat treatment with the base line welding parameters (7.2 K.amp and T=60 cycle). This test was employed to relate the metallurgical factors with the mechanical properties of the weld nugget. The etching process was achieved on the specimen using an etching solution for the weld metal, according to AWS [31].

Moreover, microscopic photographs were taken for a few zones, namely, fusion zone (cast nugget), HAZ, and the region around the HAZ with the base metal [14].

The percentage of ferrite content is measured for the welded specimen condition by using the ferrite scope to detect the ferrite content for steel welds. The calibration process was carried out on standard specimens supplied by the manufacture with the help of a sensing probe. Then after the sensing probe was passed over the test specimens which the percentage ferrite content was investigated. The ferrite content examination was carried for different zones of the spot nugget and base metal.

2.4. Determination of Weldability

The spot weldability was determined in the form of Lobe curves. It indicates the range of conditions that the satisfactory welds can be obtained. Hence, indicates the tolerance of the material to change in welding parameters according to the industrial conditions. Lobe curve can be found by systematically testing shear and tensile test. By specifying a distributed range of current values and testing welds made with varying weld times, the limit can be found and the curve established [32] as follows:

1- The upper limit of the welding current range is generally regarded as the minimum current that causes expulsion.
2- The lower limit can be established by determining [(2/3)P_{max}] at each time where (P_{max}), the maximum breaking strength can supported by the weld nugget and extending this value to intersect the line of each time, in plots of load VS current which obtained from lap and peel tests. Then this value intersecting with the current axis to obtain the minimum current from x-axis for each value of times. It was mentioned that "the $P_{breaking}$ =f(I) curves are obtained in absolutely identical natures to a point which is taken on the load VS current curves at the certain time. The first point has the joint strength equal to [(2/3)Pmax]. The second point at which there is expulsion at the end of the welding zone, [2]."

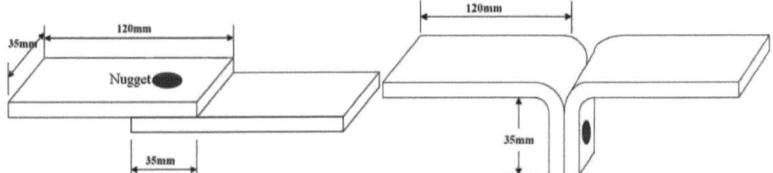

Figure 17 Geometry and the dimensions of tensile-shear test specimen

However, there are other types of specimens for the relevant service conditions see Fig. 18.

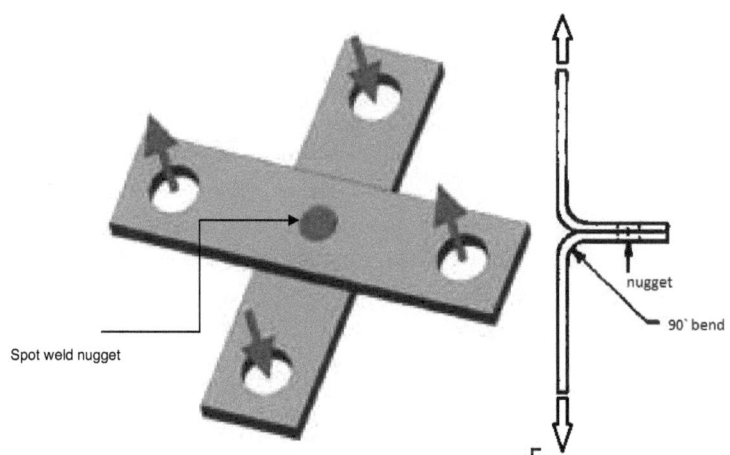

Figure 18 Cross-tension test specimens (left) [33], and peel in form of tensile test (right)

Chapter 3
Strength Evaluation of Spot Weld

3.1. Introduction

Destructive tensile tests are conducted usually in two forms of lap-specimens and peel specimens. Different welds current and time have used. The welding electrode tips have different diameters.

The lap specimens were welded under different parameters of welding times and welding currents. The maximum load and the associated nugget diameter were measured to establish Lobe curve [7]. The metallography and microhardness tests also were conducted [14].

The change in current at a constant of time will increase in heat generation (heat built up) with minimum losses in heat to the surrounding metal. This will increase in the weld area and the weld joint strength of weld joint.

The increase in current would rapidly increase in weld nugget strength, which is connected to the increasing in weld nugget area, to a critical limit of current for the spot welding of mild steel. Aidun [32] had shown that, the response of strength to increase in current is regular for the aluminum alloys spot welding.

From the heat formula the total heat developed is a linear function of time. Then, the increase in time will increase the heat losses, which are caused by conduction into the surrounding work piece and into the electrodes as well as into the surrounding air. Thus, it can be explained why the increase in weld time has a lower effect on the weld strength and weld area, as compared with the current.

The maximum load can show the difference between the lap and peel tests from the tensile test [7]. For the lap joint, the weld nugget supports the higher load when compared with the peel joint.

3.2. Spot Welding of Carbon Steel Sheet

Resistance spot welding (RSW) is the most widely common methods used for joining structures carbon steel. The desired reliability of these joints can be made. The material weldability and the reliability of these joints under different loading are investigated. The specimens of thickness 0.8 mm carbon steel number 1.8902 in a strip form were welded [7]. The strips of lap-joints and curved peel-joints configurations have been used. The welding parameters such as weld current, and weld time have been investigated to show the relation between the weld area and the joint strength properties has been presented. The obtained results were showing that the weld joint strength and the molten area (weld nugget volume) highly increase with the increasing of weld current. Therefore, the correlation between the maximum load (joint strength) and area has been given. The reliable weldability under the tensile and shearing loading was considered. Therefore, the new limits of weldabilty have been presented that consider these two types of loading. Moreover, the experimental results were compared with the empirical relations that consider the sheet thickness only. This steel has been successfully welded with a common spot-fusion process. The structural performance of the joint has been presented under tensile and shear loads.

The material employed in these testing was consisted of a single sheet of carbon cold rolled steel (16 GS, 1.8902) with a nominal thickness (0.8 mm), see [7]. No surface coating was applied. Table 2 and Table 3 show the chemical composition and the mechanical properties, respectively.

Table 2 Chemical composition of Steel % [9]

Steel Designation	C	Cr	Ni	Mn	Si	Fe
16 GS (1.8902)	0.15	0.3	0.3	2.2	0.2	Rem.

Table 3 Mechanical properties [9]

Tensile strength (MPa)	0.2% offset Yield strengths	Elongation	Reduction of area
600	365 MPa	19%	50%

However, carbon steel 1.8902 has been used in the automotive and industrial application, the weldability of carbon steel, 1.8902 has been little investigated. The ease producing of sound spot welds by the convention tensile tests has been determined in different works (Gibson, 1997, J.M. SawHill, JR., 1980, J.M. Sawhill, JR.H. Watanable, 1977, Aslanlar, Ogur, Ozsarac, Ilhan, & Demir, 2007, Kahraman, 2007, Ozyurek, 2007). In most cases, good spot-welding practice requires three parameters that have to be controlled, namely current, time and electrode pressure [35]. The increase in current to critical limits would rapidly increase in weld nugget strength, due to the increasing of the weld nugget area [9]. The effect of the atmosphere which improve the tensile-shearing strength was investigated for the RSW of titanium sheets [23]. The response of strength to increase in current is regular for the aluminum alloys spot welding [32]. Weld time is the second variable during which the current is allowed to flow. Weld time for RSW usually computed by the number of cycles as record oscillographically. The convention electronic timers can be used also by the second and converted to cycle (1 Sec=60 cycle). The optimum weld time for tensile peel and tensile shear joint of 1.2 mm galvanized chromate steel sheets was studied [36]. The variable resistance (R) is influenced by the force applied (i.e., electrode pressure) through its effect on contact resistance (Andraws, 1986, Ray & Verma, 2005). The increases of the contact resistance will increase the rate of heat generation. Therefore, the electrode pressure influences the heating by affecting the contact resistance [32]. In case the pressure automation was not available, the electrode tip diameter has the same rule.

3.2.1. Static Test

The simple static tensile test (shear-lap, tensile-peel) are widely conducted for the lap and peel specimens. The specimens were welded using an electrode tip 4 mm diameter. The effect of welding variables on the joint strength and weld area of the 1.8902 carbon steel has been categorized for both types of specimens [17]. The program includes tensile test experiments to determine the spot weldability of carbon steel under two different modes of stresses. The test procedure includes:

1- A set of peel and lap specimens were welded under different welding condition (weld time and current). The electrode tip was 4 mm.

2- A static tensile test has been made to determine the joint strength. The maximum load, in which is a major property from the design point of view, and the weld nugget area has been measured. According to the thickness, the maximum weld current 7.2 kA were applied, where the explusion can be observed. The samples joined at a lower welding current of 3.7 kA will not have an enough weld size due to the smaller heat input.

The dynamic loading test with the specific speed (strain rate) is also required. Because the application of spot welding in the automotive industry, the car body may contain of 3000-4000 spot nugget points where subjected also to the crashing [40]. Therefore, an impact tensile test is recommended to use in such cases [40]. The behavior of thin welds of different grades of steels in dynamic loading conditions has been evaluated through impact tensile test (ITT) at different temperatures. The dynamic impact test is always connected with the phenomena of transition temperature in steel which occur in a very narrow temperature range [40]. In this cases few specimens for determination the transition temperature and cracking initiating points [14]. These researches will contribute to the selection of optimal welding conditions and to the development of new grades of metals for specific applications.

3.2.2. Welding Equipments

The resistance spot welding equipment employed was a standard foot operated rocker arm spot welding (Bay Kay) with transformer capacity (15 kVA) and single phase (220 volt) [7]. Electrode was pure copper (RWMA class 1) which have a high thermal and electrical conductivity. Nevertheless, some mechanical properties are required. The dimensions were controlled by the machining process for the electrode tips. Originally, the electrodes were machined with a flat tip diameter of 4 mm, body diameter of 16 mm and the tip angles were 30ȯ truncated cones.

The electrode tip diameter was selected according to the American Welding Society (AWS) relationship Eq. (1) [3]:

$d_{el} = 0.1 + 2t$ (Eq. 1)

where d_{el}; the electrode tip diameter (in), and t is the sheet thickness (in). This relationship may not be useful for thinner thickness ranges between (0.131-0.51 mm) and for the thickness ranges between (3.18-12.7 mm) [3].

3.2.3. Welding Process

The welding process was carried out for the specimen in specific dimensions, see Figure 19. The sets of welding specimens were made at different weld variables, see Table 4. At first, the weld time and electrode force were kept constant and varying the weld current subsequently. Moreover, the electrode tip diameter was fixed. Therefore, different welds joint properties were obtained. Four welding times and currents were applied. Sixteen sets of specimens were made. Each set consist of three specimens for each time and current. Therefore, each point on the curves represents the average of three measurments. Electrode force was set to be constant at (1617 N). Hence, for the electrode diameter 4 mm, the electrode pressure was proposed to be equal to 129 MPa [7].

Table 4 The welding process parameters [9]

Weld time (cycle)	Weld current (kA)	Electrode pressure (MPa)	Electrode force (N)
15, 30, 45, 60	3.7, 5.8, 6.7, 7.2	129	1617

3.2.4. Tensile and Shearing Testing

Static strength was measured using the conventional tensile shear lap specimens and peel specimens. Specimen details and testing methods are described in AWS practice [41]. All welds were tested in tensile shear using a convention hydraulic tensile testing machine. It was equipped with a digital readout for both load and extension. The available function of the machine is to maintain both the peak load and of the weld strength, see [2].

The lap shear specimen of two flat plates of 0.8 mm thickness with an overlapping distance 25 mm was welded, see

Figure 19a. The geometry is approximated as closely as possible to pure shear made on the joint.

The peel mode specimen was subjected to a combined stress created by joining the specimen distances away from the loading plane, see

Figure 19b. The stresses were a tensile load combined with moment around the weld point. The behavior was closely approximated pure tension [7] [32]. The entire specimen dimensions were confirmed by (AWS) practice [3]. The weld defect represents the stress constraint in experimental specimens. Therefore, the effect of notch in such specimens will compensate the effect of misalignments, imperfect geometry and defects in service [40]. The desired temperature and relative conditions are determined before the static or dynamic testing.

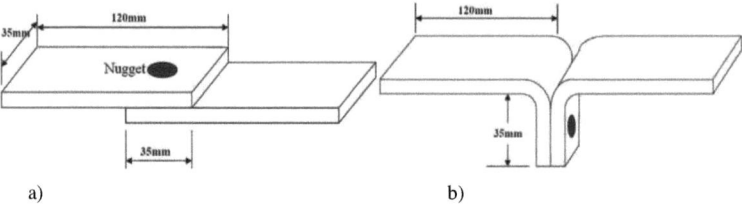

Figure 19 a) Lap-shear tensile test specimen; b) peel-tensile test specimen [7] [9] **XXX repeated !!**

3.2.5. Determination of Weldability

The spot weldability for 0.8 mm carbon steel was determined in form of Lobe curves [7]. This curve is plotted as weld time versus weld current in one form of loading. It indicates to the range of conditions in which satisfactory welds can be obtained. Lobe curve will be obtained by performing the tensile test on specimens that welded under various conditions of welding time and current by varying weld times for a specified current. The limits of the curve were established [7] [32]. The higher and lower tensile shear strength that related to the heat input amount have been demonstrated as follows [7]:

1- The upper limit of the welding current range is generally regarded as the minimum current that causes expulsion.
2- The lower limit can be established by determining (2/3 P_{max}) at each time. The maximum breaking strength can supported by the weld nugget and extending this value to intersect the x-axis of each time, in plots of load VS current which are obtained from lap and peel tests. Then this value will be projected to obtain the minimum current from x-axis for each value of times.
3- The first point on the left curve (lower bound) has joint strength (2/3 P_{max}). The second points of the right curve (upper bound) in which there is expulsion at the end of the welding zone.
4- When weld made with current and/or time exceeding the upper limit in the Lobe curve, expulsion occurs and, therefore is considered unacceptable. In contrast, welds made with current or times below the lower limit have insufficient size nuggets it will be brittle and is likewise considered unacceptable.

Because the weldability also describes the joint strength in term of the weld nugget area [15], the American Welding Society (AWS), American National Standards Institute (ANSI) and the Society of Automotive Engineers (SAE) have recommended an Eq. 2 to estimate the weld nugget area (d_n) with respect to the electrode diameter as follows:

$$d_n = 4(t)^{0.5} \tag{Eq.2}$$

Nevertheless, the fused area should have at least the following diameter [41]:

$$d_n = 0.8\,(t)^{0.5} \tag{Eq.3}$$

where t, is the thickness of the thinner of the two pieces. Therefore, the optimum nugget area for a certain electrode diameter can be estimated.

3.3. Results and Discussion

In this test, the specimen of spot weld was submitted to combined shear stresses and a certain amount of through thickness tensile stresses. The experimental results focus on the current and time effects.

3.3.1. Effect of Weld Current and Weld Time, Lap Joints

The welding current has a major influence on the tensile and shearing strength. The maximum stress was measured in which indicated to the joint strength [7]. The joint strength and the weld nugget area increase with the increasing of the weld current up to the maximum value as shown in Figure 20. Nugget growth was presented under various welding conditions [42], howevere the development of nugget was studied also by [43]. The results agree with Ref. [37]. The same behavior was obtained by increasing the weld time [7], see Figure 21. The result shown that the increasing of the weld time and electrode force increases the tensile shearing strength, i.e., weld area [23]. Kahraman et al. [23] also investigated the effect of Argon atmosphere on the tensile shearing strength.

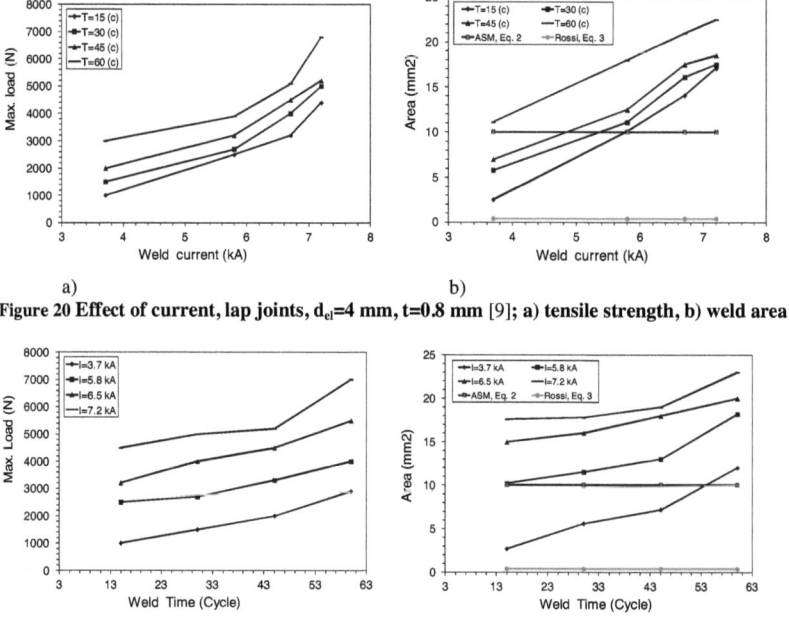

a) b)
Figure 20 Effect of current, lap joints, d_{el}=4 mm, t=0.8 mm [9]; a) tensile strength, b) weld area [7]

a) b)

Figure 21 Effect of time, lap joints [9], del=4 mm, t=0.8 mm; a) tensile strength, b) weld area [7]

A direct relationship between current, joint strength and weld area agrees with feature of the AWS curve [3]. The weld area increases rapidly with weld current, from low current levels until expulsion occurs at the higher current levels. A similar behavior occurs with the weld strength. Because the joint strength has been increased as the weld current and area are increasing. Then, the expulsion will reduce the fused metal. Hence, the enhancement in tensile shearing of the spot weld joint is attributed to the enlargements of the weld area. This result also agrees well with [35] [37] [36].

Figure 21 shows the comparison with Eq. 2 and 3. The better estimation is given from the relation of AWS. However, these linear relationships do not consider the increasing of the area due to heat generation.

Welding with electrode diameter 4 mm and low welding current (below 3.7 kA) will result in a smaller weld area [7]. Since, a shallow depth and indentation as well as smaller diameter are produced, the joint will withstand smaller load. Nevertheless, the weld area and joint strength will rapidly increase after the current of 5.8 kA. Normally, the maximum and minimum welding parameters are influenced and dependent on the product geometry, machine type, and design, electrode materials and design. Therefore, the investigation of a certain material in a production line will be needed to get these entire factors in one hand. Nevertheless, the researchers have been selected a certain couple of parameters for studying the weldablity of metals.

3.3.2. Effect of Weld Current and Weld Time, Peel Joints

Due to the configuration of this specimen, a higher tensile stress will be developed over the spot weld [7]. A significant torque arises, depending on the position of the weld point from the plain of applied load. Therefore, the tensile test results of this type of specimen show a low strength with large displacement. The weld nugget will not support the applied load. Then the failure occurs at a lower load as compared with lap joint, in which the nugget undergoes shear stress with little amount of torque around the nugget.

The weld area and joint strength increase with the welding current. For peel joints, lower values of joint strength were obtained as compared with the lap joint [7], see Figure 22. However, the area of weld nugget of the peel specimen is little higher than those measured for the lap joint.

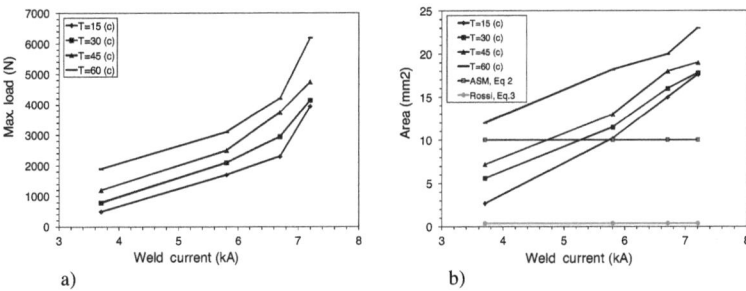

Figure 22 Effect of current, peel joints [9]; d_{el}=4 mm, t=0.8 mm; a) tensile strength, b) weld area [7]

The increasing of the welding time causes an increasing in the weld nugget area, hence increasing of the joint strength, see Figure 23.

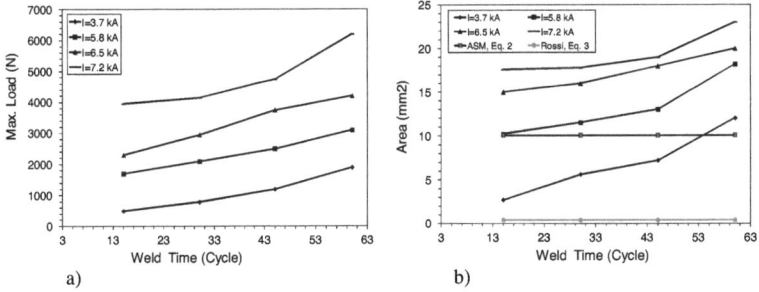

Figure 23 Effect of time, peel joints [9]; del=4 mm, t=0.8 mm; a) tensile strength, b) weld area [7]

Eq. 3 and 2 show again the lower value of weld nugget diameter according to the sheet thickness. It is to be emphasized that these lower only represent the minimum size that could be expected to carry the load for a specific thickness.

3.4. Weldability Lobe Curve

The determination of the Lobe curves has been presented [7]. Figure 24 shows the specific Lobe curve for carbon steel with nominal thickness 0.8 mm. No coating was applied for the sheet surfaces. The brittle and expulsion limits are shown. The width of the acceptable welds can be changed according to the sheet thickness, material type, heat treatments, and surface conditions.

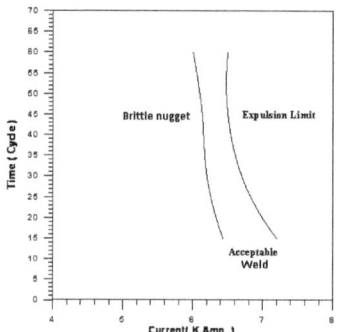

Figure 24 Lobe curve; d_{el}=4 mm, t=0.8 mm, lap joint [9]

The produced weld area and strength have a polynomial function for lap and peel joints, see Eq. (4) and (5), respectively.

$[Max.load\ (N) = 7.6609 \times Area^2 + 78.396 \times Area + 882.1]_{Shear,\ lap}$ (Eq.4)

$[Max.load\ (N) = 0.67\ [7.6609 \times Area^2 + 78.396 \times Area + 882.1]]_{Tensile,\ peel}$ (Eq.5)

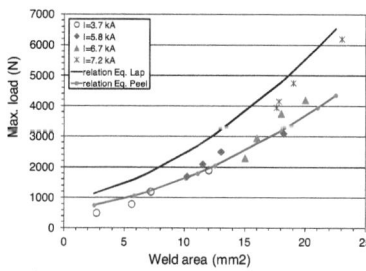

Figure 25 The relation of weld area and joint strength for lap joint and peel joints [9]

According to the upper and lower limits, the weldability has been estimated in a new vision, see Fig. 26. Therefore, the values between these two limits represent the general weldability and reliability under different loading type. The static strength has been estimated also using the linear relation as follows [40]:

$F = ktd\sigma_{bm}$ (Eq. 6)

where t is the plate thickness (mm), d the nugget diameter (mm), σ_{bm} the base metal resistance (MPa) and k is a constant (being equal to 2.5–3, with F in N) [40], see Fig. 26. The empirical relation (Eq. 8) gives the maximum design strength as based on the nugget area that influence on the joint strength. Nevertheless, the increasing in weld area is not always in a linear relation. Because it starts to be dropped due to the metallurgical and cracking effects [14].

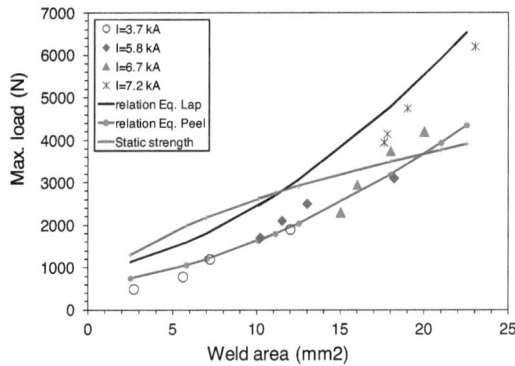

Figure 26 **The spot joint strength**

3.5. Conclusions

Carbon steel 0.8 mm has been welded [9]. Traditionally, the welding current has a major effect on the weld area and joint strength. If current is low, the weld joint will have insufficient strength and brittleness of the produced nugget. The spot welds strength depends upon the joint design. Therefore, the curved peel-joint have a lower strength as compared with the lap joint, however, the weld areas for the peel-joint is larger than those of the lap joint. A good correlation was obtained between tensile shear strength and weld nugget area. It finds that the weld strength proportion to the diameter of fused zones (weld nugget). The spot joint configuration has an effect on the heat dissipation. Therefore, the overlapped peel specimens have a higher nugget area, due to lower heat dissipation (higher heat built up) at nugget regions. The parameters that not favorable for spot reliability are determined through Lobe curve. In this work, the spot weldability has been determined under different loading conditions. Therefore, a better reliability can be determined for specific materials. The minimum weld area that needed for specific thickness was compared with the experiments results. Hence, the minimum weld area was determined as compared with AWS relation and experiments.

Chapter 4
Cracking in Spot Welded Joints

4.1. Introduction

The welded structures are subjected to cyclic loading. The major influence of such loading is concentrated in Aircraft structure where the spot welds have covered the connection areas. The model of fracture depends on the sheet properties, metallurgical and geometrical aspects. The spot welds nugget cracking of austenitic stainless steel at temperatures between 700-1010 °C was investigated [14]. Traditionally, the cracks and subsequent failures tend to occur around around the spot nugget in welded temperature [44]. Actually, these cracks are developed due to incomplete melting and inappropriate electrode pressure, which causes an expulsion of molten metal. These cracks start to grow and causing either the interface or plug fracture, according to the loading type. In this work, the micro-cracks in the weld nugget were indicated for this type of steel at elevated temperature. Cracks are appearing in a certain range of temperature about 700-750 °C. The cracks like defect and cavitations were presented. According to the fracture mechanics point of view, these cracks reduce the mechanical strength. Therefore, these cracks have to be taken into account with a certain precaution [14].

Several approaches should be employed to control phase precipitation in the weld. These approaches include: decreasing the content of ferrite, and other elements in weld [45]. Howvere, the selecting metal added weld wire during welding is recommened for other process of welding [45], in spot welding the sheel materials contents is primery considered. Nevertheless, the author recommend also to use the interfacing added sheet.

The type of the service media and conditions determine the necessity of the heat treatments. For some applications, heat treatment is used to impart the greatest degree of corrosion resistance. Sometimes, heat treatment is only necessary for stress relieving, when the service media include high temperature operations in the corrosive media. The welding process includes high melting temperature in a short time and in a small area, which is surrounded by a cooler metal. This produces a high stresses which invariably arise in any welded structure. However, the need for a stress relief treatment in welded constructed of austenitic stainless steel has been over emphasized. Therefore, heat treatments applied under the variation of welding conditions. The heat treatment reduces the mechanical properties of weld metal.

The annealing effect on the mechanical properties of the spot nugget is shown for the tensile lap-test, microhardness and metallurgical tests. This treatment has relieved approximately 90% of stress [46].

The stresses are introduced due to rapid heating, cooling and the forge pressure on the weld nugget. The effect on the mechanical properties are investigated and their effects during the metallurgical and microhardness test results [2].

The base matrix of austenite in structure with a smaller amount of ferrite is confirmed by the magnetic scope examination [2]. Because the fusion began rapidly at the interface of the two sheet surfaces and moved towards the through thickness of the sheet metal, the weld metal have a smaller amount of ferrite which vary according to heat input.

The microstructure of the as welded specimens shows that the weld metal consists of the same structure along the spot weld which consists of fully austenitic weld beads and those containing some delta ferrite which varied in amount along the spot welds. It is well known that the delta ferrite in weld in austenitic steel increases the resistance to hot cracking [47].

The maximum contents of delta ferrite was presneted at the center of spot point and decreases towards the outside. The minimum value is reached at the farthest distance from the spot weld (approximate equal to 0.18% ferrite). The amounts of the delta ferrite increase when the heat input increases with the same distribution in quantity, from the maximum at the center to the minimum towards the outside where the high heat is dissipated.

The test of the heat treated specimen, showed that the amount of the delta ferrite will be affected by the temperature of the treatment as shown in the baseline welding parameter which are 7.2 K. Amp, current and 60 cycle weld time (spec. 1). The high temperature of the annealing treatment reduced the amount of ferrite to a minimum value less than that was measured in the stress-relieved specimen and the latter contained ferrite less than that were measured in the as welded specimen as shown in Table 7.

The delta ferrite exposed for a long time to temperatures ranging from about 650-950 °C. It will transform to the intermetallic compound of chromium and iron called the sigma phase . This phase may causes loss of ductility and impact resistance, see Ref. [48]. The phenomena was observed first time in spot weld nugget in Refs [5], [14]

4.2. Welding Cracks

Austenitic stainless steel, AISI (321), cold rolled, 1.5 mm thickness without coating has been used. The overlapped specimens were welded. The chemical composition and the mechanical properties are given in **Table 5** and **Table 6**, respectively. The electrodes RWMA class-1; pure copper materials were recommended due to the high thermal and electrical conductivity [2][14][9].

Table 5 Chemical composition of stainless steel sheets [14]

Steel designation	C	Cr	Ni	Mn	V	Mo	Si	Ti	Nb	Cu	Fe
AISI 321	0.05	19	8.8	1	0.06	0.5	0.72	0.46	0.01	0.31	Rem

Table 6 Mechanical properties of stainless steel sheets [14]

Tensile strength	Yield strength (0.2% offset)	Elongation	Reduction of area
600 MPa	205 MPa	40%	50%

Welding cracks of the welded joints is considered as a serious defect. They start to grow from a certain defects until final failure. The failure tends to occur due to the crack orientation around the heat affected zone (HAZ) [49], [44], [50]. A typical through-thickness stress distribution and the fatigue critical location have been studied also at the edge of a spot weld nugget. Traditionally, the maximum stress occurs also at the interface between the two sheets [51].

A few studies deal with the cracking of the nugget area and with the ferrite contents of the austenitic stainless steel. Most studies deal with the aluminum welding, hot cracking, the welding process type, and the alloy compositions that determine the cracking susceptibility. The literature showed that the cracks initiate from the HAZ in aluminum alloy, i.e. from the periphery of spot weld nugget. The cracks were formed at elevated temperatures in the presence of liquid metal due to the metallurgical factors [21]. It was observed that the crack initiation and propagation in the weld fusion zone and the HAZ of 5083-aluminium alloy. It was found that the cracking susceptibility depended on the Magnesium contents. Therefore, Toyota reported the solidification failure in the nugget or liquation cracking in the HAZ for one of the 5000 series of aluminum alloys containing above 5 % weight of Mg [21]. The preheating will decrease the thermal stresses and the temperature gradient. Hence, the cracking ability will be decreased.

The crack propagation-based fatigue life approach for resistance spot welds was proposed [52]. Moreover, the effect of welding residual stresses was taken into account. The effects of spot weld diameter as well as the location of crack initiation have been predicted. Lin et al. [53] examined the fatigue crack paths near the spot welds in the square-cup, lap-shear and coach-peel specimens. The stress intensity factor (SIF) solutions have been used to predict the fatigue lives. SIF and fatigue lives for welding joint can be calculated using Fracture Analysis Code-2 dimension (Franc2D) for different type of materials, cracks and weld geometries [2][54][55].

However, there is an increasing use of stainless steel alloys in the industry due to their corrosion resistance and superior mechanical properties [56]; still there is a lack of practical information on their cracking due to the spot welding process. The fatigue crack appears to be initiated near the weld notch tip. Then, it will be propagated through the sheet thickness with a crack kink path angle, see Fig. 27. The crack paths have different propagation possibilities according to the maximum stress direction and the presence of porosity or solidification cracks. Hence, the crack path is changed accordingly.

An x-ray technique is used to investigate the fatigue crack initiation and crack propagation processes of spot welded joints. The cracks usually initiate 0.2-1.0 mm from the nugget edge of the spot weld at the border of the nugget [14] [57].

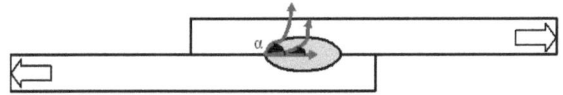

Figure 27 The proposed fatigue crack paths with a kink angle α 0-90° [14]

The effects of heating on the microstructural characteristics and the cracking ability were shown in stainless steel because at high temperature the cracks have been observed in the center of the weld nugget [14]. A specific microstructure was developed with certain ferrite content. Microhardness profiles and the weld nugget defects have been presented.

4.3. Stainless Steel Spot Weld Structure

The weld cracking and fracture toughness depending also on the weld hardness. Therefore, the microhardness distribution is investigated usually along the faying surface (longitudinal), and through the thickness (traverse), respectively, see Fig. 28. The Vickers mirohardenss is employed across the weld nugget, HAZ and the base materials using the conventional microhardness tester (JTT Digital micrometer taster, type JMT7 type A, Toshi INC.) with 300 gr loads [14]. The average of the two reading values were considered.

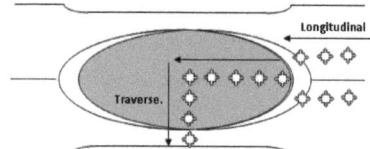

Figure 28 The microhardness measurements [14]

The percentage of ferrite content is indicated using the ferrite scope M11 instrument (Fisher manufacturing). The sensing probe passes over the specimen. The calibration process is recommended to be carried out in advance [14]. The annealing after welding produce maximum softness and ductility. suitable cooling from the annealing temperature to prevent subsequent intergranular corrosion is required [8][31]. Therefore, air-cooling is generally adequate. Annealing is performed at 1010 °C. In light section might be held at this temperature for 3 minutes per 2.5 mm. The time passed for thickness (1.5 mm) will be 2 minutes approximately. Stress relieving at temperature 750 °C for 2 min is advisable when the service environment is known to be suspected to cause stress corrosion. By using the stabilized or extra low-carbon grades, heating at stress relieve temperature could avoid the intergranular precipitates of chromium [58].

4.3.1 High Temperature Crack Growth

The stress relieving temperature exceeded 500 °C, the crack was adjacent to the weld during the post welding heat treatment and during a service at elevated temperature. At the temperature of 750°C, crack growth takes place along the interface between the austenite and delta ferrite in the weldment of 321 S.S [14].

The crack growth behaviour also observed in the weldment of 308 S.S. in the temperature range of 600 to 800°C [59]. The crack growth takes place at 600 °C along the interphace between the austenite and the arm of delta ferrite in the weldments of stainless steel. The factors influencing hot cracking of welds are substantially the same as for other steel, namely structure and impurity content. The cracks ware caused by embrittlement of weld metal and thermal stress [45]. The microstctue is primarily responsible for significant degradation in mechanical properties and microstructures of weld metal. It plays a role for crack inteoation and progation.

Weld metal that solidifies as ferrite inherently much less susceptible to cracking than that which solidifies as austenite. Mixed structures which contain more than 3% ferrite at room temperature have in practice adequate resistance to hot cracking, but the fully austenite weld metal with smaller amount of ferrite is very crack sensitive, see Ref. [14] . Therefore, after the stress relieving treatment for the spot weld metal, the weld nugget metals have (1) % ferrite at a temperature of 750 °C, Table 7. At high temperature sigma phase becomes more dominant in the crack growth process, see Ref. [14].

4.3.2 Microhardness Distributions

The microhardness for as-welded, annealed, and stress relieved specimens has been measured equal to 275, 210 and 240 HV, respectively [14]. The higher hardness distribution in longitudinal direction, see Figure 29. The hardness increases again at the center of the nugget. Since the initial contact at HAZ through a contact bridge (asperities), the asperities will be flattened (enlarged) due to the heating concentration. Therefore, the heat input increases the softening of the weld and reducing the stresses. The traverse microhardness profile has a higher value in the center of the weld nugget. Again, the microhardness drops through thickness beyond the center. The transverse hardness increases again at HAZ. This is because the compressive stresses will be developed through the thickness due to the nugget growing in a relatively cold sheet metal. The forge pressure of the electrodes balances the nugget growing force. Therefore, the compressive stresses are created around the nugget. It was found that the microstructural cracking is related with the hardness and heat input. Nevertheless, the welding process may also produce cracks. The observed nature of the hardness profiles are similar to the ones reported by [44][60]

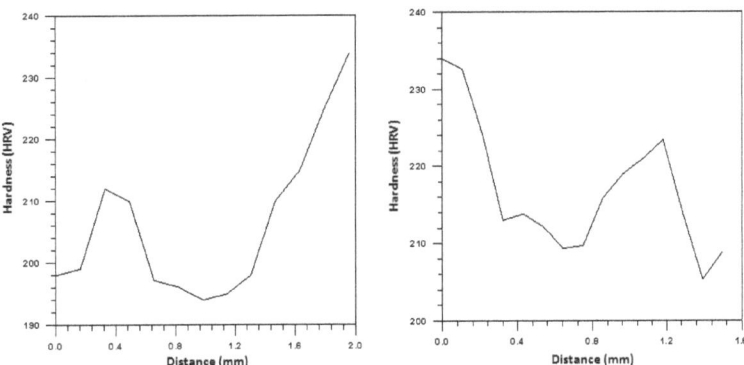

a)   As-welded conditions: (left) Longitudinal, (right) Transverse

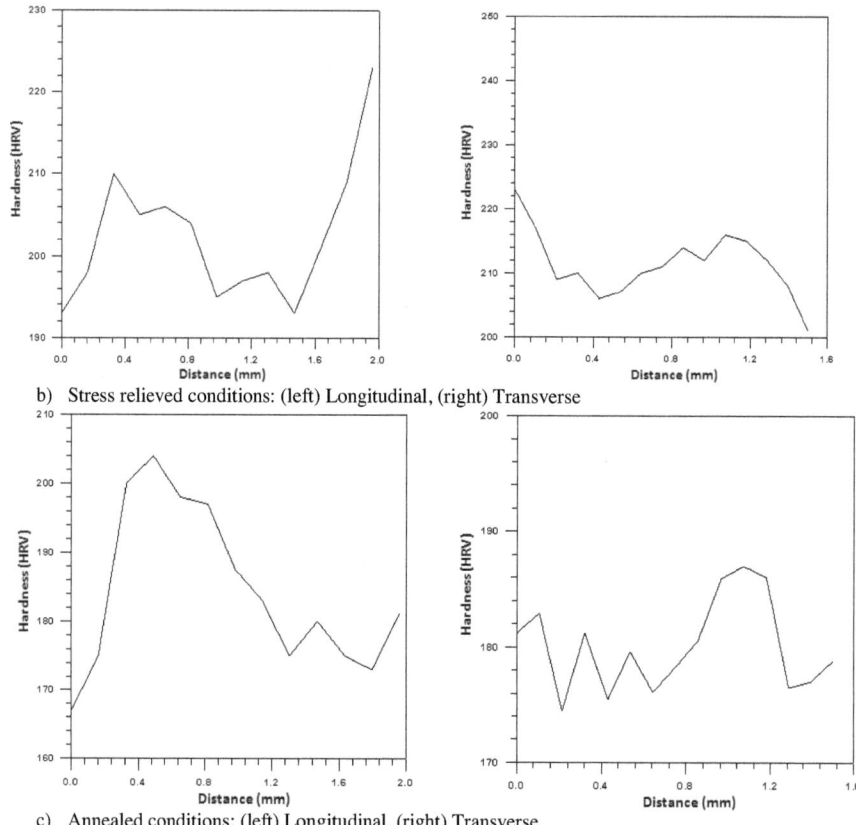

b) Stress relieved conditions: (left) Longitudinal, (right) Transverse

c) Annealed conditions: (left) Longitudinal, (right) Transverse

Figure 29 Microhardness profiles at different conditions [14]

4.3.3 High Temperature Crack Growth and Ferrite Contents

Postweld heat treatments (PWHT) reduce the residual the tensile stresses. The fracture toughness of the structures is essentially improved [24]. Heat Treatments have the effect on the mechanical properties. The type and the necessity for heat treatment of austenitic chromium-nickel steel weldments depend on the service conditions [8]. Stresses are developed due to the forge pressure and the rapid cooling of the small molten metal. Nevretheless, the post weld cold work improve the fatigue strength of Aluminuim alloy by using identing pressure [13]. The post-weld cold worked Al RSW induceses a compressive residual stresses during the cold working process [13].

These steels were annealed after welding to obtain the maximum softness and ductility. Unlike the unstablized grades, these steels did not require water quenching or other acceleration of cooling from the annealing temperature to prevent subsequent intergranular corrosion, air-cooling is generally adequate [8]. Annealing of austenitic stainless steel was performed at 1010°C. In light section might be held at this temperature for 3 minutes per 2.5 *mm*. The time passed for thickness (1.5 *mm*) will be 2 min approximately then the weldments are cooled in air.

Stress relieving employed at the temperature 750 °C. The holding time is depending on the thickness of the specimens. The rapid heating and cooling in spot welding processes for stainless steel has been overemphasized. Therefore, stress relieving generally advisable when the service environment is known or suspected to cause stress corrosion [14]. By using stabilized or extra low-carbon grades, heating at stress relieve temperature could avoid the intergranular precipitates of chromium.

The etching solution of 10 ml acetic acid, 15 ml hydrochloric acid (HCL), 10 ml nitric acid (HNO3), and two drops of glycerol was used for the spot weld area according to the AWS [8]. The cracks have been appearing around the periphery of the spot nugget due to the stress concentration and the notch effect in HAZ. This is the most common type of failure observed in tensile and shear spot welded samples occurred by cracks which propagated circumferentially around the weld nugget [40] [61].

The multi-site cracking around the weld nugget and the expelled fused metal are shown in

Figure 30 , respectively. Fracture mechanics with help of FE have been used to estimate the crack length and path in other welded joints [55], [62].

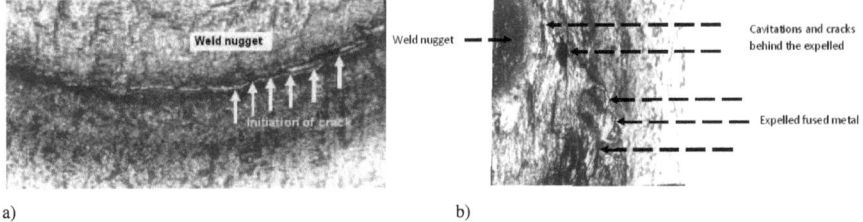

a) b)

Figure 30 Spot weld cracking [14]; a) crack initiation around the weld nugget, b) the expelled weld metal [14]

The contacting point metal will be melted and poured during the process, Figure 31 . The irregular shape cavity which produced after the solidification, see Figure 31 . The irregular shape cavity will initiate the crack that propagates through the weld metal. The crack will be extended during the loading, or during the electrodes removing that were sticking to the surfaces as illustrated in Figure 27. This crack is also related to the electrode pressure, cooling rate, and the fused metal depth between the two sheets. Weld metal that solidifies as ferrite inherently much less susceptible to cracking than that which solidifies as austenite [63]. The mixed structures which solidify with ferrite contents more than 3% at room temperature have in practice adequate resistance to hot cracking. Because certain amount of delta ferrite (5–8%) will scavenge the impurities and take care of a certain amount of thermal strain [47].

However, the austenite weld metal with smaller amount of ferrite is very crack sensitive. The crack growth occurred in the weldments of austenitic stainless steel at the temperature of stress relieving treatment of 750 °C with 1% ferrite content, see Figure32 . Table 7 shows the ferrite percentage in different conditions.

Table 7 Delta-ferrite contents [14]

Spec. No.	I (kA)	T (cycle)	Delta-ferrite Amount from the center of the spot weld (%)					
1 (As-welded)	7.2	60	1.4	1.2	1	0.8	0.24	0.2
2 (As-welded)	5.8	45	0.9	0.76	0.67	0.23	0.22	0.2
3 (As-welded)	3.7	30	0.43	0.27	0.25	0.24	0.2	0.2
4 (annealed)	7.2	60	0.24	0.22	0.21	0.2	0.2	0.2
5 (stress relieved)	7.2	60	1	0.9	0.8	0.65	0.2	0.2

Kamaraj et al. [64] studied the crack growth of un-welded stainless steel. They have found that the crack growth takes place at 600 °C along the interface between the austenite and the arm of delta ferrite in the weldments of stainless steel type 308. In this work, the stress relieving temperature is exceeded 500 °C and will develop about 1 % of ferrite at a temperature of 750 °C, see Table 7. The specimens were finally annealed at 1010 °C which produce the reliable microstructure with a minimum number of defects and about 0.24 % of ferrite, see Figure 33.

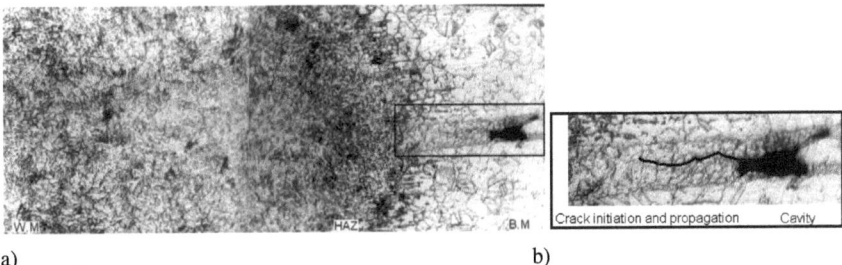

a) b)
Figure 31 As-welded structures, I=7.2 kA, T=60 cycles; a) and b) cracking and the cavity in the solidified metal around HAZ [14]

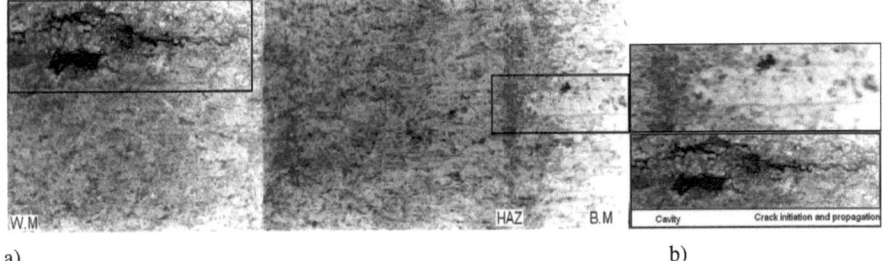

a) b)
Figure32 Stress-relieved structures, I=7.2 kA, T=60 cycles; a) cracking of the weld nugget area, b) the interface molten metal [14]

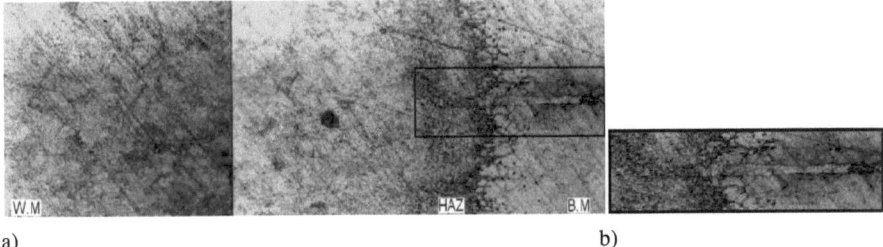

a) b)
Figure 33 Annealed structures, I=7.2 kA, T=60 cycle, a) and b) weld area and the interface molten metal [14]

4.3. Spot Welding Defects

Occasionally, numbers of individual defects appear in the spot weld nugget. The large cavity is the most conventional defects. It reduces the joint quality and fatigue strength. Figure 34 shows the expulsion cavity and the surrounding small shrinkage cavity that occurs at high machine setting of stainless steel [14]. Practically, most welds have a shrinkage cavity in the center of the weld nugget [65]. The finite element simulation could be used to determine the regions of stress concentration which in turn determine the crack initiating and fracture strength [66]. Nevrthless, the initial cracks and cavity in range of 0.1 mm was expected in welds [55].

A cavity which occurs from the heavy expulsion of molten metal may extend over a part in a few millimeters. The metal shrinkage porosity and cavity will extend and accelerate the crack propagation. The high welding power setting (i.e., high welding current and time) causes a high speed movement of the molten metal. Therefore, an opposite force against the electrodes forge direction will be developed. Hence, an expelled metal will be extruded once the nugget force increases the supporting force. The electrode force should be sufficient to balance the compressive stresses that developed within the nugget. Moreover, welding settings have to be compatible with the sheet thickness. It is to be emphasized that only the molten metal in the hot zone at the center of the spot nugget will be expelled out. Splashing in spot welding reduces the cross section being welded. Figure 34 (b) shows the fractured specimens and the expelled molten metal from the hot zone.

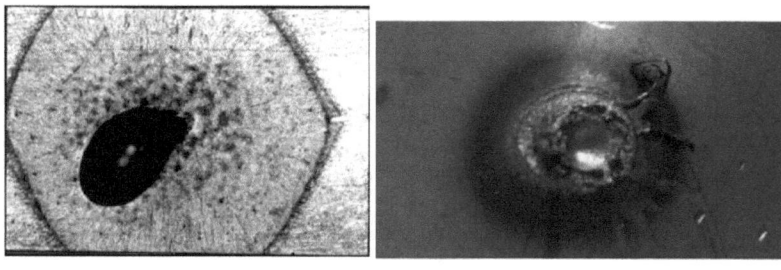

a) b)
Figure 34 a) cavitations due to the metal expulsion (Etching: 10 ml acetic acid, 15 ml hydrochloric acid, 10 ml nitric acid and 2 drops of glycerol); b) the expelled of plastic metal from the hot zone [14]

The crack start from the center of porosity hole due to the low electrode froce, see Fig. 35 (left). The crack from the edge and propagating through the sheet thickness [67].

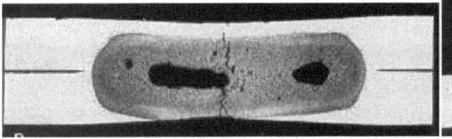 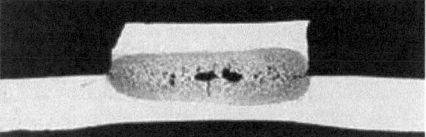

Figure 35 Porosity and discontinuity in aluminum alloy (left), The fracture from the edge nugget (right) [67]

4.4. Conclusions

In spot welding, the HAZ and weld nugget area have a critical role. The cracking phenomena have been investigated in spot-welded joint. The cracking ability is a major problem in welded structures. This is because the cracks reduce the joint strength. Nevertheless, literatures are limited about high temperature cracking at the weld nugget of resistance spot welded stainless steel alloys. The cracking behavior at high temperature in this part was observed. It is concluded that the cracks initiate from the center of sport weld nugget at the temperature up to 750 °C where the amount of produced ferrite about 1% [14]. The crack growth will take place also along the interface between the austenite and delta ferrite. A certain precaution must be taken by considering the temperature limits and by reducing the element that may develop ferrite particles. However, the crack may also be developed at the welding conditions from HAZ around the weld nugget due to notch stress concentration. If the electrode pressure is not sufficient to balance the developed force within the spot due to the nugget growth, part of the weld metal will be expelled out. The expulsion of the highly molten metal from the hot zone at the center of the spot will increase the stresses nears the point where the loss of metal occurs. Actually, the cracks around the weld nugget can be developed also due to the electrodes sticking to the surface. By removing the electrodes and the weld nugget still ductile, the crack will initiate due to the sticking force and grows toward the nugget center.

References

[1] P. Iwe and M. J. Greitmann, "Welding through the Ages resistance welding DVS - German Welding Society International Institute of Welding (IIW) Expert of Commission III ,, Resistance welding , solid state welding and allied joining processes " Greitmann Consulting , Contents Welding," pp. 1–75, 2013.

[2] A. M. Al-Mukhtar, "Spot Welding Efficiency and It ' S Effect on Structural Strength of Gas Generator and Its Performance," Baghdad University, 2002.

[3] A. W. Society, *Welding*. 1958.

[4] M. P. Groover, *Fundamentals of modern manufacturing: materials processes, and systems*. John Wiley & Sons, 2007.

[5] A. M. Al-Mukhtar, "Spot Welding Efficiency and its Effect on Structural Strength of Gas Generator and its Performance," Baghdad University, 2002.

[6] T. Uwaba, Y. Yano, and M. Ito, "Resistance spot weldability of 11Cr–ferritic/martensitic steel sheets," *J. Nucl. Mater.*, vol. 421, no. 1–3, pp. 132–139, Feb. 2012.

[7] A. M. Al-Mukhtar and Q. Doos, "The Spot Weldability of Carbon Steel Sheet," *Adv. Mater. Sci. Eng.*, vol. 2013, pp. 1–6, 2013.

[8] *American Welding Society*. 1985.

[9] A. M. Q. D. Al-Mukhtar and Q. Doos, "The Spot Weldability of Carbon Steel Sheet," *Adv. Mater. Sci. Eng.*, vol. 2013, pp. 1–6, 2013.

[10] J. S. Jr, H. Watanabe, and J. Mitchell, "Spot Weldability of Mn-Mo-Nb, VN and SAE 1008 Steels," *Weld. J.*, pp. 217–224, 1977.

[11] M. Benachour, M. Benguediab, A. Hadjoui, F. Hadjoui, and N. Benachour, "Fatigue crack growth of a double fillet weld," *Comput. Mater. Sci.*, vol. 44, pp. 489–495, 2008.

[12] Y. J. Chao, "Failure mode of spot welds: interfacial versus pullout," *Sci. Technol. Weld. Join.*, vol. 8, no. 2, pp. 133–137, 2003.

[13] D. Kim, D. Blake, S. J. Ryu, and B. S. Lim, "A Study on Fatigue Strength Improvement of Aluminum Alloy Resistant Spot Welds by Cold Working," *Mater. Sci. Forum*, vol. 539–543, pp. 3961–3966, 2007.

[14] A. Al-Mukhtar and Q. Doos, "Cracking Phenomenon in Spot Welded Joints of Austenitic Stainless Steel," *Mater. Sci. Appl.*, vol. 4, no. October, pp. 656–662, 2013.

[15] H. Moshayedi and I. Sattari-Far, "Numerical and experimental study of nugget size growth in resistance spot welding of austenitic stainless steels," *J. Mater. Process. Technol.*, vol. 212, no. 2, pp. 347–354, Feb. 2012.

[16] T. Triyono, J. Jamasri, M. N. Ilman, and R. Soekrisno, "Fatigue Behavior of Resistance Spot-Welded Unequal Sheet Thickness Austenitic Stainless Steel," *Mod. Appl. Sci.*, vol. 6, no. 5, pp. 34–42, Apr. 2012.

[17] S. Aslanlar, "Welding time effect on mechanical properties of automotive sheets in electrical resistance spot welding," 2008.

[18] I. Sevim, "Effect of hardness to fracture toughness for spot welded steel sheets," *Mater. Des.*, vol. 27, no. 1, pp. 21–30, Jan. 2006.

[19] V. Balasubramanian, V. Ravisankar, and G. Madhusudhan Reddy, "Effect of pulsed current welding on mechanical properties of high strength aluminum alloy," *Int. J. Adv. Manuf. Technol.*, vol. 36, no. 3–4, pp. 254–262, Jan. 2007.

[20] D. K. A. and R.W.bennett, "Effect of resistance welding variables on the strength of spot-welded 6061-t6. Aluminum alloy," *Weld. J.*, vol. 64, no. 12, p. 6061, 1980.

[21] J. Senkara and H. Zhang, "Cracking in Spot Welding Aluminum Alloy AA5754," *Weld. Reseach*, no. July, 2000.

[22] a. M. Pereira, J. M. Ferreira, a. Loureiro, J. D. M. Costa, and P. J. Bártolo, "Effect of process parameters on the strength of resistance spot welds in 6082-T6 aluminium alloy," *Mater. Des.*, vol. 31, no. 5, pp. 2454–2463, May 2010.

[23] N. Kahraman, "The influence of welding parameters on the joint strength of resistance spot-welded titanium sheets," *Mater. Des.*, vol. 28, no. 2, pp. 420–427, Jan. 2007.

[24] V. . Acoff, R. . Thompson, R. . Griffin, and B. Radhakrishnan, "Effect of heat treatment on microstructure and microhardness of spot welds in Ti□26Al□11Nb," *Mater. Sci. Eng. A*, vol. 152, no. 1–2, pp. 304–309, May 1992.

[25] A. M. Al-Mukhtar, "Investigation of Some Welding Parameters in Resistance Spot Welding of Austenitic Stainless," in *5th Sc. Conference of the College of Engineering, Baghdad University*, 2003.

[26] T. Kim, Y. S. Lee, J. Lee, and S. H. Rhee, "A Study of Nondestructive Weld Quality Inspection and Estimation during Resistance Spot Welding," *Key Eng. Mater.*, vol. 270–273, pp. 2338–2344, 2004.

[27] T. A. Welder, "Welding Aluminum Wire feeder Technology The American Welder We ' ve Got," *Quality*, no. April, 2006.

[28] T. Triyono, Y. Purwaningrum, and I. Chamid, "Critical Nugget Diameter of Resistance Spot Welded Stiffened Thin Plate Structure," *Mod. Appl. Sci.*, vol. 7, no. 7, pp. 17–22, 2013.

[29] and J. C. B. J.M. SawHill, JR., "Spot Weldability of High-Strength Sheet Steels," *Weld. J.*, p. 1980, 1980.

[30] T. K. Pal and K. Chattopadhyay, "Resistance spot weldability and high cycle fatigue behaviour of martensitic (M190) steel sheet," *Fatigue Fract. Eng. Mater. Struct.*, vol. 34, no. 1, pp. 46–52, Jan. 2011.

[31] A. W. Society, *Welding Handbook*. 1958.

[32] D. K. Aidun and R. W. Bennett, "Effect of resistance welding variables on the strength of spot welded 6061-T6 aluminum alloy," 1985.

[33] E. Preliminar and M. De Soldas, "Preliminary Study on the Mechanical Behavior of Friction Spot Welds," *Soldag. inp. São Paulo*, vol. 14, no. 3, pp. 238–247, 2009.

[34] S. W. Gibson, "Advanced Welding," 1997.

[35] and J. W. M. J.M. Sawhill, JR.H. Watanable, "Spot Weldability of Mn Mo Cb, V-N and SAE 1008 Steels," 1977.

[36] S. Aslanlar, a. Ogur, U. Ozsarac, E. Ilhan, and Z. Demir, "Effect of welding current on mechanical properties of galvanized chromided steel sheets in electrical resistance spot welding," *Mater. Des.*, vol. 28, no. 1, pp. 2–7, Jan. 2007.

[37] D. Ozyurek, "Materials & Design An effect of weld current and weld atmosphere on the resistance spot weldability of 304L austenitic stainless steel," *J. Mater.*, 2007.

[38] D. R. Andraws, "The importance of monitoring resistance welding parameters," vol. 54, no. April, p. 1986, 1986.

[39] P. K. Ray and B. B. Verma, "A study on spot heating induced fatigue crack growth retardation," *Engineering*, vol. 28, pp. 579–585, 2005.

[40] E. Bayraktar, D. Kaplan, and M. Grumbach, "Application of impact tensile testing to spot welded sheets," *J. Mater. Process. Technol.*, vol. 153–154, pp. 80–86, Nov. 2004.

[41] B. E. Rossi, *Welding Engineering*. 1985.

[42] B.-W. Cha and S.-J. Na, "A study on the relationship between welding conditions and residual stress of resistance spot welded 304-type stainless steels," *J. od Manuf. Syst.*, vol. 22, no. 3, pp. 181–189, 2003.

[43] J. E. Gould, "An examination of nugget development during spot-welding, using both experimental and analytical techniques," *Weld. J.*, vol. 66, no. 1, pp. S1–S10, 1987.

[44] G. Mukhopadhyay, S. Bhattacharya, and K. K. Ray, "Strength assessment of spot-welded sheets of interstitial free steels," *J. Mater. Process. Technol.*, vol. 209, no. 4, pp. 1995–2007, Feb. 2009.

[45] K. S. Guan, X. D. Xu, Y. Y. Zhang, and Z. W. Wang, "Cracks and precipitate phases in 321 stainless steel weld of flue gas pipe," *Eng. Fail. Anal.*, vol. 12, no. 4, pp. 623–633, Aug. 2005.

[46] J. American, "JOURNALS AMERICAN WELDING," *Society*.

[47] K. Guan, X. Xu, H. Xu, and Z. Wang, "Effect of aging at 700°C on precipitation and toughness of AISI 321 and AISI 347 austenitic stainless steel welds," *Nucl. Eng. Des.*, vol. 235, no. 23, pp. 2485–2494, Dec. 2005.

[48] A. INTERNATIONAL, *Welding, brazing and soldering*, vol. 6. ASM International, 1993.

[49] M. Vural and A. Akkus, "On the resistance spot weldability of galvanized interstitial free steel sheets with austenitic stainless steel sheets," *J. Mater. Process. Technol.*, vol. 153–154, pp. 1–6, Nov. 2004.

[50] H. Yu, S. Yang, H. Kang, H. Kim, and K. Kim, "Fatigue Life Analysis of Spot Weldment of Cold Rolled and High Strength Steel Using FEM," *Trans. Korean Soc. …*, 2008.

[51] M. M. Rahman, "Fatigue Life Prediction of Spot-Welded Structures: A Finite Element Analysis Approach," *Eur. J. Sci. Res.*, vol. 22, no. 3, pp. 444–456, 2008.

[52] S. E. Mirsalehi and a. H. Kokabi, "Fatigue life estimation of spot welds using a crack propagation-based method with consideration of residual stresses effect," *Mater. Sci. Eng. A*, vol. 527, no. 23, pp. 6359–6363, Sep. 2010.

[53] S. Lin, J. Pan, P. Wung, and J. Chiang, "A fatigue crack growth model for spot welds under cyclic loading conditions," *Int. J. Fatigue*, vol. 28, no. 7, pp. 792–803, Jul. 2006.

[54] A. M. Al-Mukhtar, H. Biermann, P. Hübner, and S. Henkel, "Comparison of the Stress Intensity Factor of Load-Carrying Cruciform Welded Joints with Different Geometries.pdf," *J. Mater. Eng. Perform.*, 2010.

[55] A. M. Al-Mukhtar, H. Biermann, P. Hübner, and S. Henkel, "Determination of Some Parameters for Fatigue Life in Welded Joints Using Fracture Mechanics Method," *J. Mater. Eng. Perform.*, vol. 19, no. 9, pp. 1225–1234, Mar. 2010.

[56] D. Özyürek, "An effect of weld current and weld atmosphere on the resistance spot weldability of 304L austenitic stainless steel," *Mater. Des.*, vol. 29, no. 3, pp. 597–603, Jan. 2008.

[57] G. Wang and M. E. Barkey, "Investigating the Spot Weld Fatigue Crack Growth Process Using X-ray Imaging," *Weld. J.*, vol. 85, no. April, p. 84s–90s, 2006.

[58] and V. I. R. A. I. Pugachev, N. B. Demkin, "Dimensions of Initial Contact in Spot Welding of Light Alloys, Welding Production .," no. 4, pp. pp13–15, 1968.

[59] M. Kamaraj and V. M. Radhakrishnan, "High temperature crack growth in austenitic weld metal," *Eng. Fract. Mech.*, vol. 33, no. 5, pp. 801–811, Jan. 1989.

[60] B.-H. Choi, D.-H. Joo, and S.-H. Song, "Observation and prediction of fatigue behavior of spot welded joints with triple thin steel plates under tensile-shear loading," *Int. J. Fatigue*, vol. 29, no. 4, pp. 620–627, 2007.

[61] S.-H. Lin, J. Pan, S.-R. Wu, T. Tyan, and P. Wung, "Failure loads of spot welds under combined opening and shear static loading conditions," *Int. J. Solids Struct.*, vol. 39, no. 1, pp. 19–39, 2002.

[62] A. Al-Mukhtar and H. Biermann, "The effect of weld profile and geometries of butt weld joints on fatigue life under cyclic tensile loading," *J. Mater. ...*, p. 23, 2011.

[63] "J.F. Lancaster, The Metallurgy of Welding, Brazing and Soldering, London, George Alden and Unwin LTD, 1970.," p. 1970, 1970.

[64] and V. M. R. M. Kamaraj, "High Temperature Crack Growth in Austenitic Weld," vol. 33, no. 5, p. 1989, 1989.

[65] A. M. Al-Mukhtar, "Static Strength Behavior of Austenitic Stainless Steel Sheet, Journal of Engineering College, Vol. 10, No. 2, June 2004, Baghdad University.," *J. Eng. Coll.*, vol. 10, no. 2, p. 2004, 2004.

[66] A. M. Al-Mukhtar, "Investigation of the thickness effect on the fatigue strength calculation," *J. Fail. Anal. Prev.*, p. 23, 2013.

[67] a Gean, S. a Westgate, J. C. Kucza, and J. C. Ehrstrom, "Static and Fatigue Behavior of Spot-Welded 51 82-0 Aluminum Alloy Sheet," *Weld. Journal- New York*, vol. 78, no. March, p. 80s–86s, 1999.

Index

A

acceptable welding, 5
alualloy, 15
aluminum, 14, 17, 23, 24, 35
austenitic, 15, 33, 34, 35, 36, 38, 39

B

brazing, 6
Burrs, 18

C

carbon steel, 17, 24
cavity, 39, 41, 42
compressive, 12, 37, 38, 42
conductivity, 9, 12, 13, 25, 34
corrosiin, 15
corrosion, 15, 33, 35, 36, 38, 39
Corrosion, 16
corrosive media, 15, 33
crack growth, 5, 36, 37, 39, 41, 43
cracking, 5, 15, 25, 31, 33, 34, 35, 36, 37, 39, 41, 43
cracks, 33, 34, 35, 36, 37, 39, 42, 43

D

defects, 5, 8, 12, 17, 26, 34, 36, 41, 42
ductility, 19, 34, 36, 38

E

electrode, 9, 11, 12, 13, 14, 15, 17, 23, 24, 25, 27, 28, 33, 39, 42, 43
experiments, 5, 24, 32
expulsion, 18, 19, 21, 27, 28, 30, 33, 42, 43

F

failure, 15, 19, 29, 34, 35, 39
fatigue, 12, 34, 35, 38, 42
ferrite, 5, 21, 33, 34, 35, 36, 39, 40, 41, 43
forging, 6
Fracture, 35, 39

H

heat, 5, 6, 7, 8, 9, 12, 13, 15, 17, 20, 23, 25, 27, 28, 30, 31, 33, 34, 36, 37, 38
heat treatments, 5, 30, 33, 38

J

joining, 6, 7, 8, 14, 18, 23, 26
joining process, 6
joint geometry, 10
joint strength, 17, 21, 23, 24, 25, 27, 28, 29, 30, 31, 43

L

lap joint, 23, 29, 30, 31
lap-joints, 23
Lobe curve, 14, 17, 21, 23, 26, 27, 30, 31
Lobe curves, 21, 26, 30

M

material, 6, 7, 12, 13, 14, 20, 21, 24, 29, 30
mechanical properties, 5, 12, 15, 17, 20, 24, 25, 33, 34, 35, 38
microhardness, 20, 21, 23, 33, 36, 37
Microhardness, 36, 37, 38
Microstructure, 36
molten weld, 19

N

nugget, 5, 9, 12, 18, 19, 21, 23, 24, 25, 27, 29, 31, 33, 35, 36, 37, 39, 42, 43
nugget center, 43

P

peel joint, 23
peel joints, 29, 30
pullout, 19

R

reliability, 9, 23, 31
resistance, 6, 9, 10, 12, 13, 14, 15, 18, 24, 25, 31, 33, 34, 35, 37, 39

S

shear, 18, 19, 23, 24, 26, 27, 29, 31, 35, 39
sheet thickness, 7, 23, 25, 30, 35, 42
shrinkage, 42
softening, 5, 37
soldering, 6
spot weldability, 5, 14, 15, 17, 21, 24, 26, 31
stainless steel, 15, 36, 41
steel, 15, 17, 19, 21, 23, 24, 25, 26, 30, 31, 33, 34, 35, 36, 38, 39, 41, 42
Steel, 15, 23, 24, 34
sticking force, 43
strength, 5, 7, 8, 12, 14, 15, 19, 21, 23, 24, 26, 27, 28, 29, 30, 31, 33, 34, 38, 42
Stress, 36, 38, 39, 41
stress relieving, 5, 15, 33, 36, 37, 39, 41
surface conditions, 12, 13, 30

T

temperature, 5, 8, 12, 15, 25, 26, 33, 34, 35, 36, 38, 39, 41, 43
tensile, 17, 18, 19, 21, 23, 24, 25, 26, 27, 28, 29, 31, 33, 38, 39

thickness, 7, 13, 14, 23, 24, 25, 26, 27, 30, 31, 34, 36, 37, 39

W

weld area, 5, 13, 15, 17, 23, 24, 28, 29, 30, 31, 39, 41
weld nugget, 9, 12, 17, 18, 19, 21, 23, 24, 25, 27, 29, 30, 31, 33, 35, 36, 37, 39, 41, 42, 43
weld strength, 15, 17, 19, 23, 24, 26, 28, 31
welding condition, 19, 20, 24
welding conditions, 14, 17, 25, 33, 43
welding current, 9, 10, 12, 21, 25, 27, 28, 29, 31, 42
welding technology, 6
welding time, 9, 14, 17, 26, 29
welding times, 14, 23, 25

Z

zone, 21, 27, 34, 35, 42, 43

I want morebooks!

Buy your books fast and straightforward online - at one of the world's fastest growing online book stores! Environmentally sound due to Print-on-Demand technologies.

Buy your books online at

www.get-morebooks.com

Kaufen Sie Ihre Bücher schnell und unkompliziert online – auf einer der am schnellsten wachsenden Buchhandelsplattformen weltweit!
Dank Print-On-Demand umwelt- und ressourcenschonend produziert.

Bücher schneller online kaufen

www.morebooks.de

SIA OmniScriptum Publishing
Brivibas gatve 197
LV-103 9 Riga, Latvia
Telefax: +371 68620455

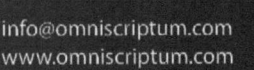
info@omniscriptum.com
www.omniscriptum.com

Printed by Books on Demand GmbH, Norderstedt / Germany